TABLE OF CONTENTS

TOPIC	PAGE #
Introduction	2
A Simple Method For Making Conversions	3
Commonly Used SI (metric) Units and Symbols	6
Examples of Correct and Incorrect SI (metric) Symbol Usage	7
Conversion Symbols	8
Decimal Equivalents of Common Fractions	9
Legend of Symbols	10
Linear Measure (length)	10
Mass	14
Area (length x width)	18
Water and Flow	21
Volume (liquid) (length x width x height)	23
Volume (dry)	30
Density of Liquids, Solids, and Soil	32
Temperature	34
Luminance; and Luminous Intensity and Flux	34
Work, Power, and Energy; or Heat Work	36
Dynamic and Kinematic Viscosity	41
Pressure	41
Torque or Moment of Force	44
Velocity and Speed (distance ÷ time)	44
Acceleration (velocity ÷ time)	46
Force (mass x acceleration)	47
Frequency; Angular Frequency; and Angular Velocity	48
Electric Current and Charge	48
SI (metric) Multiples for Hertz (Hz)	49
Plane Angles	50
Time (a quick reference)	50
More Time Definitions and Conversions	52
Parts Per Million (ppm)	55
Radioactivity	57
Miscellaneous	57
Some Common Formulas	58
Rates of Application	60
Potassium and Phosphorus	61
Approximate Measures	61

CONVERSION FACTORS AND DATA FOR THE LIFE AND PHYSICAL SCIENCES

INTRODUCTION

While with Cal Poly, San Luis Obispo, and Colorado State University I needed a complete listing of common conversion factors and other helpful information for my students. But what was available was incomplete and lacking at best. Plus, nearly all conversion tables and charts were extremely messy, chaotic, and NEVER user friendly. And so early on I started a compilation of my own. Today, that "compilation" has grown into a 60+ page booklet, and it has been extremely helpful and useful through the years.

So this "compilation" is primarily a book of conversion factors for helping with transposing from the International System of Units (or metric system) to the English system of measurement (and vice versa). But other conversions and units are also listed when relevant, or of general interest to the reader (or to the author).

This publication also contains other useful information including helpful common formulas; various parts per million specifics; chemistry info; plus other helpful information for anyone involved in the life and physical sciences (chemistry, biology, agriculture, horticulture, geology, physics, and other life and physical sciences).

It would be great if we lived in a world where we didn't need conversion factors... but we don't. And converting from one system of measurement to another can be tiresome, and just a general pain. Of course, anyone can go the internet and find answers to most conversions, but being able to have a hard copy can be the quickest and simplest way to find answers. Plus a book can be taken to class, to the job place, or anywhere away from internet access. And so that is why this book has been compiled together.

While every effort has gone into making this book 100% accurate and error-free, there is bound to be oversights (hopefully extremely minor) and omissions. So, if the reader would like to contribute to the next edition, please feel free to contact me so those additions will be included. Now. I know this book will make your lives easier. You can thank me later.

Brent Rouppet, Ph.D.
soildoctor@fertilesoilsolutions.com
www.fertilesoilsolutions.com

A SIMPLE METHOD FOR MAKING CONVERSIONS:
Using Conversion Factors to Switch From the English System of Measurement to the International System of Units (SI) [metric system], and Vice Versa

Some people are "scared stiff" of the International System of Units [or metric system]; others have trouble (at best) converting back and forth from one system to the other. On the other hand, some folks are just plain annoyed with any conversions and converting. None-the-less, we are stuck with different systems of measurement, and must from time to time make conversions. However when the following examples are followed, converting from one system to the other can not only be quite easy, it can be fun [for some obsessive/compulsive (i.e., sick) folks, anyway]. But first, some background.

The International System of Units (SI) is a modernized version of the metric system established by international agreement. The metric system of measurement was developed during the French Revolution and was first promoted in the U.S. by Thomas Jefferson, believe it or not. Its use was legalized (was this necessary?) in the U.S. in 1866. In 1902, proposed congressional legislation requiring the U.S. Government to use the metric system exclusively was defeated by a single vote.

The International System of Units provides a logical and interconnected framework for all measurements in science, industry, and commerce. And the metric system is much simpler to use than the existing English system since all it units of measurement are divisible by 10.

So let's make several conversions from the English system of measurement to the International System of Units:

1. Un easy example: A hay grower has just reported an alfalfa yield of 5.0 tons/acre. However, he needs to know what his cutting yielded in metric tonnes (megagrams)/ hectare. ♪: For mental reference, one hectare is approximately the size of a baseball diamond; while one acre is approximately the size of an American football field.

From the "area" section of the following conversion factors, we find that one acre ≡ 0.4047 hectare [we also find one hectare ≡ 2.471 acres]. Next, in the "mass" section of the conversion factors, we find that 1 U.S. short ton [or 2000 pounds] ≡ 0.90718 metric tonne [we also find one metric tonne (megagram) ≡ 2204.6 pounds or 1.1023 tons]. We set the problem up as shown below:

$$\frac{5.0 \text{ tons hay}}{1 \text{ acre}} \times \frac{0.90718 \text{ tonne}}{1 \text{ ton}} \times \frac{1 \text{ acre}}{0.4047 \text{ hectare}} \equiv 11.21 \text{ tonnes/hectare}$$

We multiply across the top: (5.0 x 0.90718 x 1 ≡ 4.54); and then do the same on the bottom: (1 x 1 x 0.4047 ≡ 0.4047). Now we simply divide the top answer by the bottom answer and we get the tonnes/hectare that we are looking for:

$$\frac{4.54}{0.4047} \equiv 11.21 \text{ tonnes/hectare}$$

Not only is this method simple, it can be fun (well, sort of...).

♫: Whatever you are converting "from" must ALWAYS cancel out, leaving what you are converting "to." In this case, <u>tons and acres</u> MUST cancel themselves out, leaving <u>tonnes and hectares</u>.

We could have just as easily obtained the same answer by setting up the problem as follows, while getting the same end result...it is only a matter of preference for the individual doing the math:

$$\frac{5.0 \text{ tons hay}}{1 \text{ acre}} \times \frac{1 \text{ tonne}}{1.1023 \text{ tons}} \times \frac{2.471 \text{ acres}}{1 \text{ hectare}} \equiv 11.21 \text{ tonnes/} 1 \text{ hectare}$$

2. <u>A more challenging example</u>: For silliness sake, let's find out how many millimeters it is from Earth to the Sun. This problem will take an extra step, but remember, whatever we are converting "from" must ALWAYS cancel out, leaving what we are converting "to." We know that the Sun is just about 92,893,864 miles from Earth. From the "linear measure" section of the following conversion factors, we find that 1 kilometer ≡ 0.6214 miles, and 1 mile ≡ 1.609 kilometers [we also find that 1 kilometer ≡ 1000 meters; and 1 meter ≡ 1000 millimeters]. So we will set up the problem as follows:

$$\frac{92{,}893{,}864 \text{ miles}}{\text{Earth to sun}} \times \frac{1 \text{ kilometer}}{0.6214 \text{ miles}} \times \frac{1000 \text{ meters}}{1 \text{ kilometer}} \times \frac{1000 \text{ millimeters}}{1 \text{ meter}}$$

≈ 1.49 × 10^{14} millimeters from Earth to the Sun

That is a lot of millimeters…

Again, we could have just as easily used 1 mile ≡ 1.609 kilometers instead of 1 kilometer ≡ 0.6214. It's simply a matter of preference for the individual doing the math:

3. <u>One more (kind of) silly problem</u>: Let's calculate how many milliliters there are in 1 acre-foot of water. From the "volume" section we find that there are 325,850.6 gallons per acre-foot; 3.7854 liters in 1 gallon; and 1000 milliliters per liter. So the problem is set up as follows:

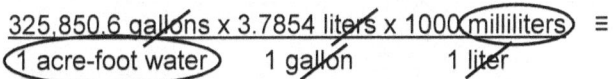

1 233 474 861 milliliters/
1 acre-foot water

♪: Just remember that whatever you are converting "from" must **ALWAYS** cancel out, leaving what you are converting "to." In this case, <u>gallons and liters</u> **MUST** cancel themselves out, leaving <u>acre-feet and milliliters</u>.

COMMONLY USED SI (METRIC) UNITS AND SYMBOLS

Type of Measurement	Unit Name	Symbol
length; width; thickness; girth; distance; etc.	millimeter centimeter meter kilometer	mm cm m km
mass (weight)	microgram milligram gram kilogram megagram (1 metric tonne) ♪: megagram, not metric tonne, is the SI preferred and accepted base unit	µg mg g kg Mg; t
time	second hour	s h
temperature	kelvin (SI base unit) degrees Celsius	K (not °K) °C
area	square meter hectare square kilometer	m^2 ha km^2
velocity	meters per second	m/s
velocity (vehicle speed, etc.)	kilometers per hour	km/h
volume	milliliter cubic centimeter liter cubic meter	mL (or ml) cm^3 L (or l) m^3
power	watt kilowatt	W kW
energy	joule kilojoule megajoule kilowatt hour	J kJ MJ kW·h
electric current	ampere	A

density	kilograms per cubic meter	kg/m³
force	newton	N
luminous intensity	candela	cd
pressure; stress	kilopascal	kPa
substance	mole	mol

EXAMPLES OF CORRECT AND INCORRECT SI (METRIC) SYMBOL USAGE

Unit/Term	Correct Usage	Incorrect Usage
kilometer	km	Km, km., KM, kms, K, k
meter	m	M, m.
millimeter	mm	Mm, mm., MM
liter	L or l	L., l.
milliliter	mL or ml	ML, Ml, mL., ml., mls
kilogram	kg	KG, KG., Kg, Kg., kgr, kgs, kilo
gram	g	G, G., g., gr, gm, GR, GM, GRM, grms
microgram	µg	mcg
hour	h	hr, hrs, HR, h., HR., HRS.
second	s	sec, S, SEC, sec., s., S.
cubic centimeter	cm³	cc, CC
kilometer per hour	km/h	KPH, kph, kmph, km/hr
kilohertz	kHz	KHz, KHZ, Khz
megahertz	MHz	MHZ, Mhz
hectopascal	hPa	HPa, HPA, Hpa, mb
kilopascal	kPa	KPa, KPA, Kpa
degree Celsius[1]	°C	C, deg C, ° C, C°
degree Fahrenheit	°F	F, deg F, ° F, F°
kelvin	K	°K, deg K, degrees k

[1] ♪: The temperature unit kelvin (K) correctly has no degree [°] sign. However, the non-SI symbols for Celsius [°C] and Fahrenheit [°F] have degree signs in order to avoid confusion with coulomb [C] and farad [F]. Also, the term "centigrade" is obsolete. The proper term is Celsius.

CONVERSION SYMBOLS
PREFIXES AND SYMBOLS LISTED BELOW ARE COMMONLY USED TO FORM NAMES AND SYMBOLS OF THE DECIMAL MULTIPLES AND SUB-MULTIPLES OF THE SI (METRIC) UNITS

Multiplication Factor	Prefix
$1,000,000,000,000,000,000,000,000 \equiv 10^{24}$	yotta (Y) (septillion)
$1,000,000,000,000,000,000,000 \equiv 10^{21}$	zetta (Z) (sextillion)
$1,000,000,000,000,000,000 \equiv 10^{18}$	exa (E) (quintillion)
$1,000,000,000,000,000 \equiv 10^{15}$	peta (P) (quadrillion)
$1,000,000,000,000 \equiv 10^{12}$	tera (T) (trillion)
$1,000,000,000 \equiv 10^{9}$	giga (G)(billion)
$1,000,000 \equiv 10^{6}$	mega (M) (million)
$1,000 \equiv 10^{3}$	kilo (k) (thousand)
$100 \equiv 10^{2}$	hecto (hecta) (h) (hundred)
10	deca (da) (ten)
0	one
$0.1 \equiv 10^{-1}$	deci (d) (tenth)
$0.01 \equiv 10^{-2}$	centi (c) (hundredth)
$0.001 \equiv 10^{-3}$	milli (m) (thousandth)
$0.000\,001 \equiv 10^{-6}$	micro (μ) (millionth)
$0.000\,000\,001 \equiv 10^{-9}$	nano (n) (billionth)
$0.000\,000\,000\,001 \equiv 10^{-12}$	pico p) (trillionth)
$0.000\,000\,000\,000\,001 \equiv 10^{-15}$	femto (f) (quadrillionth)
$0.000\,000\,000\,000\,000\,001 \equiv 10^{-18}$	atto (a) (quintillionth)
$0.000\,000\,000\,000\,000\,000\,001 \equiv 10^{-21}$	zepto (z) (sextillionth)
$0.000\,000\,000\,000\,000\,000\,000\,001 \equiv 10^{-24}$	yocto (y) (septillionth)

DECIMAL EQUIVALENTS OF COMMON FRACTIONS

1/2	.5000		3/11	.2727
1/3	.3333		4/5	.8000
1/4	.2500		4/7	.5714
1/5	.2000		4/9	.4444
1/6	.1667		4/11	.3636
1/7	.1429		5/6	.8333
1/8	.1250		5/7	.7143
1/9	.1111		5/8	.6250
1/10	.1000		5/9	.5556
1/11	.0909		5/11	.4545
1/12	.0833		5/12	.4167
1/16	.0625		6/7	.8574
1/32	.0313		6/11	.5455
1/64	.0156		7/8	.8750
2/3	.6667		7/9	.7778
2/5	.4000		7/10	.7000
2/7	.2857		7/11	.6364
2/9	.2222		7/12	.5833
2/11	.1818		8/9	.8889
3/4	.7500		8/11	.7273
3/5	.6000		9/10	.9000
3/7	.4286		9/11	.8182
3/8	.3750		10/11	.9091
3/10	.3000		11/12	.9167

LEGEND OF SYMBOLS

Symbol	Definition
≡	exactly equal to
=	equal to
≈	approximately equal to
(H)	primarily of historical interest

LINEAR MEASURE (LENGTH)

1 kilometer (km)	≡ 0.6214 statute mile ≡ 1093.6 yards ≡ 3280.8 feet ≡ 39 370 inches ≡ 10 hectometers ≡ 1000 meters ≡ 1 000 000 millimeters
1 hectometer (hm)	≡ 100 meters ≡ 10 dekameters
1 dekameter (dam)	≡ 10.936 132 983 4 yards ≡ 10 meters
1 meter (m)	≡ 1.093 613 298 yards ≡ 3.280 83 feet ≡ 39.37 inches ≡ 1000 millimeters ≡ 100 centimeters ≡ 10 decimeters ≡ 1.0×10^6 micrometers (μm) [or microns]
1 decimeter (dm)	≡ 3.937 inches ≡ 100 millimeters
1 centimeter (cm) ♪: For reference, a CD or DVD is 12 centimeters (120 millimeters) across. The diameter of the center hole is 15 millimeters	≡ 0.3937 inch ≡ 0.032 81 feet ≡ 0.0109 yard ≡ 0.01 meter ≡ 10 millimeters ≡ 10 000 micrometers (μm) [or microns] ≡ 1.0×10^8 angstroms [Å] ≡ 393.7 mils

1 millimeter (mm)	≡ 0.039 37 inch ≡ 1000 micrometers (μm) [or microns] ≡ 1.0 x 10^7 angstroms [Å]
1 micrometer (μm) [or micron]	≡ 0.000 039 37 inch ≡ 1 x 10^{-6} meter ≡ 0.001 millimeter ≡ 10 000 angstroms [Å]
1 nanometer (nm) (millimicron)	≡ 0.000 000 039 37 inch ≡ 0.001 micrometer (μm) [or micron]
1 angstrom [Å]	≡ 1 x 10^{-10} meter ≡ 0.1 nanometer
1 fermi (fm)	≡ 1 x 10^{-15} meter
1 Planck length (the base unit in the system of Planck units)	≡ 16.162 x 10^{-36} meter ≡ 16.162 x 10^{-27} nanometer
1 statute (or land) mile (mi) (international)	≡ 1.609 344 kilometers ≡ 1609 meters ≡ 1760 yards ≡ 5280 feet ≡ 63 360 inches ≡ 80 chains (or gunters) ≡ 8 furlongs ≡ 320 rods ≡ 8000 links
1 mile (U.S. survey)	= 5280 feet ≡ 1 609.347 219 meters (which is about 3.219 mm (1/8 inch) longer than the international mile. The international mile is exactly 0.0002% less than the U.S. survey mile
1 yard (yd) (international)	≡ 0.9144 meter ≡ 0.009 144 hectometer (hm) ≡ 3 feet ≡ 36 inches
1 quarter	≡ 1/4 yard (that makes sense) ≡ 0.2286 meter
1 pace	≡ 2.5 feet ≡ 0.762 meter

1 foot (ft) (international)	≡ 0.3048 meter ≡ 30.48 centimeters ≡ 0.999998 U.S. survey foot ≡ 12 inches ≡ 0.333 33 (or 1/3) yard ≡ 0.060 606 rod
1 foot (ft) (U.S. survey)	= 12 inches ≡ 1200/3937 meter = 0.304 800 609 601 250 meter
1 inch (in) international	≡ 2.540 centimeters ≡ 0.0254 meter ≡ 0.083 33 feet ≡ 1/36 yard ≡ 1/12 foot
1 barleycorn (H)	≡ 1/3 inch ≈ 8.46×10^{-3} meter
1 mil; thou (mil)	≡ 0.001 inch ≡ 2.54×10^{-5} meter ≡ 0.0254 millimeter
1 span	≡ 9 inches ≡ 0.2286 meter
1 hand	≡ 4 inches ≡ 0.1016 meter
1 palm	≡ 3 inches ≡ 0.0762 meter
1 finger	≡ 7/8 inch ≡ 0.022 225 meter
1 finger (cloth)	≡ 4-1/2 inches ≡ 0.1143 meter
1 nail (cloth)	≡ 2-1/4 inches ≡ 0.057 15 meter
1 stick (H)	≡ 2 inches ≡ 0.0508 meter
1 vara	≡ 33 inches in the United States of America. [however, vara "varies" in other countries from 32-43 inches.]
1 fathom (fm)	≡ 6 feet ≡ 1.8288 meters
1 rod (rd) (or pole, or perch)	≡ 5.5 yards ≡ 16.5 feet

	≡ 25 links
	≡ 5.0292 meters
1 link (lnk) (gunter's, surveyor's)	≡ 7.92 inches
	≡ 1/100 chain
	≡ 0.66 foot
	≡ 0.201 168 meter
1 chain (ch) or gunter (surveyor's)	≡ 100 links
	≡ 4 rods
	≡ 66 feet
	≡ 792 inches
	≡ 20.116 84 meters
1 rope (H)	≡ 20 feet
	≡ 6.096 meter
1 cable length (imperial)	≡ 608 feet
	≡ 185.3184 meters
1 cable length (international)	≡ 1/10 nautical mile
	≡ 185.2 meters
1 cable length (U.S.)	≡ 720 feet
	≡ 120 fathoms
	≡ 219.456 meters
1 furlong (fur)	≡ 40 rods
	≡ 10 chains or gunters (surveyors)
	≡ 660 feet
	≡ 220 yards
	≡ 1/8 statute mile
	≡ 201.168 meters
80 chains (ch)	≡ 1 statue mile
	≡ 320 rods
	≡ 5280 feet
	≡ 1.609 344 kilometers
1 league (lea) [land]	≡ 3 U.S. statute miles
	≡ 4.828 0417 kilometers
1 league (nl) [nautical]	≡ 3 nautical miles
	≡ 5.556 kilometers
	≡ 5556 meters
1 nautical mile (nmi) (international)	≡ 1.150 779 statute miles
	≡ 1.852 kilometers
	≡ 1852 meters
	≡ 0.999 U.S. nautical mile
	≡ 6076.115 49 feet

1 fathom	≡ 6 feet ≡ 1.8288 meters
1 bolt (U.S., cloth)	≡ 40 yards ≡ 36.576 meters
astronomical unit (AU) (approximate distance from earth to sun--varies with perihelion to aphelion)	≈ 92 893 864 miles ≈ 1.495 x 10^8 kilometers
1 light-year (ly) [distance light travels in a vacuum in 365.25 days]	≈ 5 865 696 000 000 miles = 9 460 730 472 5808 kilometers
1 parsec	≈ 3.26 light years
1 exameter	≈ 110 light years
cubit (H) [distance from fingers to elbow]	≈ 18 inches ≈ 0.5 meter
ell (H)	≡ 45 inches (in England) ≡ 1.143 meters
1 mickey	≡ 1/200 inch ≡ 1.27 x 10^{-4} meter
1 twip (twp)	≡ 1/1440 inch ≡ 1.7638 x 10^{-5} meter ≡ 17.638 micrometers (μm) [or microns]

MASS

1 pound (lb av) (avoirdupois) ♪: avoirdupois (pronounced av·oir·du·poiz) is a system of weights used in many English-speaking countries. It is based on the pound, which contains 16 ounces or 7000 grains. 100 pounds (U.S.) or 112 pounds (imperial) is equal to 1	≡ 0.453 592 37 kilogram ≡ 453.592 grams ≡ 16 ounces ≡ 7000 grains ≡ 2.731 597 339 06 x 10^{26} atomic mass units (u)

hundredweight and 20 hundredweights equals 1 ton	
1 pound (metric)	≡ 500 grams
1 pound (lb t) (troy)	≡ 5760 grains ≡ 0.373 241 7216 kilogram
1 ounce (oz av) (avoirdupois)	≡ 0.028 349 523 125 kilogram ≡ 28.349 523 125 grams ≡ 0.0625 pound ≡ 1/16 pound ≡ 16 drams (avoirdupois) ≡ 437.5 grains
1 ounce (oz t) (apothecary; troy)	≡ 1/12 pound (troy) ≡ 31.103 4768 grams
1 ounce (oz) U.S. food nutrition labeling	≡ 28 grams
1 pennyweight (dwt; pwt)	≡ 1/20 ounce (troy) ≡ 1.555 173 84 grams
1 point	≡ 1/100 carat = 2 milligrams
1 ton (sh tn) (short)	≡ 0.907 184 74 megagram (Mg) (metric tonne) ≡ 907.184 74 kilograms ≡ 2000 pounds
1 ton (long tn, or ton) (long)	≡ 2240 pounds ≡ 1016.046 9088 kilograms
1 metric tonne (megagram) (Mg) ♪: megagram, not metric tonne, is the SI preferred and accepted term	≡ 1000 kilograms ≡ 2204.6 pounds ≡ 1.1023 short tons
100 pounds	≡ 45.359 237 kilograms ≡ 1 hundredweight (short cwt.) ≡ 0.4536 quintal
1 hundredweight (sh cwt) (short); cental	≡ 100 pounds (avoirdupois) = 45.359 237 kilograms
1 hundredweight (cwt) (long)	≡ 112 pounds (avoirdupois) = 50.802 345 44 kilograms

1 stone (st)	≡ 14 pounds (avoirdupois) = 6.350 293 18 kilograms
1 quarter (informal)	≡ 1/4 short ton ≡ 226.796 185 kilograms
1 quarter, long (informal)	≡ 1/4 long ton ≡ 254.011 7272 kilograms
1 quarter (imperial)	≡ 1/4 long hundredweight = 28 pounds (avoirdupois) = 12.700 586 36 kilograms
1 kip	≡ 1000 pounds (avoirdupois) ≡ 453.592 37 kilograms
1 slug; geepound; hyl (slug) (an imperial mass unit)	≡ 32.2 pounds ≈ 14.593 903 kilograms
1 megagram (Mg) (metric tonne) ♪: megagram, not metric tonne, is the SI preferred and accepted term	≡ 1000 kilograms ≡ 2204.6 pounds ≡ 1.1023 tons
1 quintal (q) (metric)	≡ 100 kilograms ≡ 220.5 pounds
1 kilogram (kg) ♪: 1 liter of pure water at 4 °C and one standard atmosphere pressure weighs exactly 1 kilogram	≡ 1000 grams ≡ 2.2046 pounds ≡ 0.0685 slug (an imperial mass unit) ≡ 9.81 newtons (N)
1 grave (G) (the original name of the kilogram)	≡ 1 kilogram
1 centigram (cg)	≡ 10 milligrams
1 decigram (dg)	≡ 10 centigrams ≡ 100 milligrams
1 dekagram (dag)	≡ 10 grams
1 hectogram (hg)	≡ 10 dekagrams ≡ 100 grams
1 gram (g)	≡ 0.035 27 ounce ≡ 981 dynes

♪: For reference, a U.S. nickel weighs about 5 grams; a U.S. penny about 2.5 grams; one piece of M & M candy (red preferably) weighs ≈ 0.86 gram (860 mg)	≡ 15.432 grains ≡ 0.001 kilogram ≡ 1000 milligrams ≡ 1.0 x 10⁶ micrograms ♪: 1 milliliter of pure water at 4 °C and one standard atmosphere pressure weighs exactly 1 gram
1 milligram (mg)	≡ 1000 micrograms ≡ 0.0154 grain ≡ 2.2046 x 10⁻⁶ pound ≡ 0.000 035 27 ounce
1 microgram (μg)	≡ 0.000 001 gram
1 atomic mass unit (u; AMU)	≈ 1.660 538 x 10⁻²⁷ kilogram
1 electronvolt (eV)	= 1.7826 x 10⁻³⁶ kilogram
1 electron mass	≡ 9.11 x 10⁻³¹ kilogram ≡ 9.11 x 10⁻²⁸ gram ≡ 5.46 x 10⁻⁴ atomic mass unit (u)
1 proton mass	≡ 1.007 28 atomic mass units (u)
1 neutron mass	≡ 1.00866 atomic mass units (u)
1 grain ♪: the usage of grain originated from one grain of wheat; and is the smallest unit of mass in the system of weights used in the United States of America)	= 1/7000 pound (average) = 64.798 91 milligrams = 0.064 79 gram
1 sheet	≡ 1/700 pound (avoirdupois) ≡ 647.9891 milligrams
1 scruple (s ap) (apothecary)	≡ 20 grains ≡ 1.295 9782 grams
1 mite	= 1/20 grain (apothecaries) = 3.239 9455 milligrams
1 mite (metric)	≡ 1/20 gram ≡ 50 milligrams

1 dram (avoirdupois)	≡ 27.344 grains ≡ 1.771 845 195 3125 grams
1 dram (apothecary; troy)	≡ 60 grains ≡ 3.887 9346 grams
1 carat (kt)	≡ 3.086 grains ≈ 205.196 548 333 milligrams
1 carat (ct) (metric)	≡ 200 milligrams
1 point	≡ 0.01 carat ≡ 2 milligrams
1 crith (the weight of a liter of hydrogen at 0° C and 760 millimeter pressure)	≈ 89 960 milligrams ≈ 0.0896 gram ≈ 1.38274 grains
1 pennyweight	≡ 1.555 grams
1 bag (coffee)	≡ 60 kilograms
1 mark	≡ 8 ounces (troy) ≡ 248.827 8144 grams
1 clove	= 8 pounds (average) = 3.628 738 96 kilograms
1 bag (Portland cement)	= 94 pounds = 42.637 682 78 kilograms
1 wey (H)	≡ 252 pounds ≡ 18 stones ≡ 114.305 277 24 kilograms
1 barge	≡ 22-1/2 short tons ≡ 20 411.656 65 kilograms
pounds per acre x 1.12	≡ kilograms per hectare
kilograms per hectare x 0.893	≡ pounds per acre

AREA (length x width)

1 township	≡ 36 square miles (6 miles square) ≡ 36 sections ≡ 93.239 571 972 1 square kilometers

1 square mile	≡ 1 section ≡ 640 acres ≡ 27 878 400 square feet ≡ 2.59 square kilometers ≡ 258.999 hectares
1 square mi (sq mi) (US survey)	≡ 640 acres ≡ 2.589 998 47 × 10^6 square meters
1 section (1 mile × 1 mile)	≡ 1 square mile ≡ 2.589 988 110 336 × 10^6 square meters
1 barony	≡ 4000 acres ≡ 1.618 742 × 10^7 square meters
¼ section	≡ 160 acres
1 yardland	≡ 30 acres ≡ 1.2 × 105 square meters
1 acre (ac) (international) (208.71035 feet square) [i.e., 208.71035 feet × 208.71035 feet] ♪: For reference, an acre is the approximate size of an American football field	≡ 1 chain (66 feet) × 1 furlong (660 feet) ≡ 0.4046 856 4224 hectare ≡ 40.46 856 4224 ares ≡ 4046.856 4224 square meters ≡ 0.004 046 856 square kilometer ≡ 43 560 square feet ≡ 4840 square yards (1 chain × 10 chains) ≡ 0.001 562 5 square mile ≡ 160 square rods ≡ 10 square chains (gunters) ≡ 100 000 square links
1 acre (ac) (U.S. survey)	≡ 10 square chains ≡ 4840 square yards ≡ 43 560 square feet ≡ 4046.873 square meters (compare to international acre above)
1 rood (ro)	≡ 1/4 acre ≡ 1011.714 1056 square meters
1 square chain (sq ch) (international)	≡ 66 feet × 66 feet ≡ 1/10 acre ≡ 404.685 642 24 square meter

1 square chain (sq ch) (US survey)	≡ 66 feet x 66 feet ≡ 1/10 acre ≡ 404.6873 square meters
100 square feet	≡ 1/435.6 acre (or 0.002 296 acre) ≡ 9.29 square meters
1 square yard (sq yd) (1 yard x 1 yard)	≡ 0.000 206 6 acre ≡ 9 square feet ≡ 1296 square inches ≡ 0.836 127 36 square meter
1 square foot (sq ft) (1 foot x 1 foot)	≡ 144 square inches ≡ 0.111 11 square yard ≡ 0.092 903 04 square meter ≡ 929.0304 square centimeters
1 square foot (sq ft) (US survey)	≡ 144 square inches ≡ 0.092 903 411 613 2749 square meter
1 square inch (sq in)	≡ 0.006 94 square feet ≡ 6.4516 square centimeters ≡ 645.16 square millimeters
1 circular inch (circ in) [$A = \pi r^2$]	≡ 0.785 398 square inch ≡ $5.067\ 075 \times 10^{-4}$ square meter
1 square rod (or square pole or square perch)	≡ 0.006 25 acre ≡ 30.25 square yards ≡ 272.25 square feet ≡ 25.293 square meters
1 square chain	≡ 16 square rods
1 square link (sq lnk) (gunter's international)	≡ 1 link x 1 link ≡ 0.66 feet x 0.66 feet ≡ 0.4356 square feet ≡ $4.046\ 856\ 4224 \times 10^{-2}$ square meter
1 square link (sq lnk) (US survey)	≡ 0.4356 square feet ≡ $4.046\ 872 \times 10^{-2}$ square meter
1 guntha	≡ 121 square yards ≡ 101.17 square meters
1 board	≡ 1 inch x 1 foot ≡ $7.741\ 92 \times 10^{-3}$ square meter
1 square kilometer (km^2)	≡ 100 hectares ≡ 1.0×10^6 square meters ≡ 247.105 acres

1 hectare (ha) (100 meters square) [i.e., 100 meters x 100 meters] ♪: For reference, a hectare is the approximate size of a baseball diamond	≡ 0.3861 square mile ≡ 2.471 acres ≡ 11 960 square yards ≡ 0.003 85 square mile ≡ 10 000 square meters ≡ 100 ares
1 dunam (or stremma)	≡ 1000 square meters
1 are (a) (metric system)	≡ 0.000 1 square kilometer ≡ 0.01 hectare ≡ 100 square meters ≡ 1076.44 square feet ≡ 119.599 square yards ≡ 0.000 038 6 square mile ≡ 0.025 acre
1 square meter (centare)	≡ 1.196 square yards ≡ 10.7643 square feet ≡ 1.0×10^{20} square angstroms [Å]
1 square decimeter	≡ 15.5 square inches
1 square centimeter	≡ 100 square millimeters ≡ 0.1550 square inch
1 square millimeter	≡ 0.002 square inch
1 barn	≡ 10^{-28} square meter

WATER and FLOW

1 cubic foot per second (ft^3/s)	= 448.83 gallons per minute = 26 929.8 U.S. gallons per hour = 0.0826 acre feet per hour = 1 acre-inch in 1 hour and 30 seconds (or 0.992 acre-inch per hour) = 1 acre-foot in 12 hours and 6 minutes = 1.984 acre-feet per 24 hours = 0.028 316 846 592 cubic meter

	per second
1 cubic foot per minute (ft^3/m)	= 4.719 474 432 x 10^{-4} cubic meter per second
1 cubic inch per minute (in^3/min)	= 2.731 1773 x 10^{-7} cubic meter per second
1 cubic inch per second (in^3/s)	= 1.638 7064 x 10^{-5} cubic meter per second
1 gallon per minute (GPM) (U.S. fluid)	= 0.002 23 cubic feet per second = 1 acre inch in 452.6 hours = 0.002 21 acre-inch per hour = 1 acre-foot in 226.3 days = 0.004 42 acre-foot per day = 1 inch depth of water over 96.3 square feet in 1 hour = 6.309 019 64 x 10^{-5} cubic meter per second
1 gallon per hour (GPH) (U.S. fluid)	= 1.051 503 273 x 10^{-6} cubic meter per second
1 gallon per day (GPD) (U.S. fluid)	= 4.381 263 638 x 10^{-8} cubic meter per second
1 cubic meter per second (m^3/s) (SI base unit)	= 1000 liters per second ≈ 157-1/2 cubic fathoms per second = 264.172 051 US gallons per second = 35.314 454 cubic feet per second ≈ 1.305 cubic yards per second = 86.4 megaliters per day = 25 566.497 acre-feet per year = 31 536 000 cubic meters per year = 1 113 676 621 cubic feet per year = 0.007 565 84 cubic mile per year
1 cubic yard per minute	= 0.45 cubic foot per second = 3.367 gallons per second = 12.74 liters per second
1 liter per minute (LPM)	= 1.6 x 10^{-5} cubic meter per second
1 million gallons per day	= 1.547 cubic feet per second = 694.4 gallons per minute

VOLUME (LIQUID) [length x width x height]

1 acre-foot (ac ft)	≡ 43 560 cubic feet ≡ 12 acre-inches ≡ 325 851.431 889 U.S. gallons ≡ 1 233 481.837 547 52 liters ≡ 1233.481 837 547 cubic meters
1 acre-inch (ac in)	≡ 3630 cubic feet ≡ 27 154 U.S. gallons ≡ 1/12 acre-foot ≡ 102 790.153 128 96 liters ≡ 102.790 153 128 9 cubic meters
1 acre-furrow-slice soil (top six inches)	≡ 21 780 cubic feet
1 cubic mile (cu mi) [1 mile x 1 mile x 1 mile]	≡ 4 168 181 825.440 cubic meters
1 cubic yard (of fluid)	≡ 27 cubic feet ≡ 46 656 cubic inches ≡ 201.97 U.S. gallons ≡ 1615.9 U.S. pints fluid ≡ 764.554 857 984 liters ≡ 0.764 554 857 984 cubic meter ≡ 764 554 857 cubic centimeters
1 cubic foot (cu ft) (of fluid)	≡ 7.4805 gallons ≡ 1728 cubic inches ≡ 0.037 04 cubic yard ≡ 29.92 U.S. fluid quarts ≡ 28.317 liters ≡ 0.028 316 846 592 cubic meter ≡ 0.8036 bushel (dry) ≡ 59.84 U.S. pints ≡ 62.428 pounds ≡ 28.32 kilograms
1 cubic inch (cu in) (of fluid)	≡ 0.5541 fluid ounce ≡ 4.433 fluid drams ≡ 16.387 064 cubic centimeters ≡ 16.387 milliliters ≡ 0.000 578 7 cubic foot ≡ 0.004 329 gallon ≡ 0.016 387 064 liter

1 barrel (fl bl US) (U.S. fluid)	≡ 31.5 U.S. gallons ≡ 0.119 240 471 196 cubic meter
1 barrel (bl imp) (imperial)	≡ 36 gallons (imperial) ≡ 0.163 659 24 cubic meter
1 barrel (bl) (petroleum)	≡ 42 U.S. gallons ≡ 0.158 987 294 928 cubic meter
1 bucket (bkt) (imperial)	≡ 4 gallons (imperial) ≡ 0.018 184 36 cubic meter
1 peck (pk) (imperial)	≡ 2 gallons (imperial) ≡ $9.092\ 18 \times 10^{-3}$ cubic meter
1 hogshead (hhd imp) imperial	≡ 63 U.S. gallons ≡ 2 barrels (imperial) ≡ 0.327 318 48 cubic meter
1 hogshead (hhd U.S.) (U.S.)	≡ 2 fluid bushels (U.S.) ≡ 0.238 480 942 392 cubic meter
1 cubic meter (m^3) (SI unit) ♪: 1 cubic meter of pure water at 4 °C and one standard atmosphere pressure weighs exactly 1 megagram (1 metric tonne)	≡ 1000 liters ≡ 1.0×10^6 cubic centimeters ≡ 1.308 cubic yards ≡ 35.31 cubic feet ≡ 61 023 cubic inches ≡ 264.2 U.S. gallons ≡ 2113 U.S. pints fluid ≡ 28.38 bushels (dry)
1 liter (L) ♪: 1 liter of pure water at 4 °C and one standard atmosphere pressure weighs exactly 1 kilogram	≡ 1000 milliliters ≡ 1 cubic decimeter ≡ 1000 cubic centimeters ≡ 0.001 cubic meter ≡ 0.2642 U.S. gallon ≡ 1.057 fluid quarts ≡ 33.81 fluid ounces ≡ 0.22 imperial gallon ≡ 0.0353 cubic foot ≡ 61.024 cubic inches
1 cubic centimeter (also 1 milliliter) ♪: 1 cubic centimeter of pure water at 4 °C and one standard atmosphere pressure weighs exactly 1 gram	≡ 0.000 001 cubic meter ≡ 0.001 liter ≡ 0.061 cubic inch ≡ 0.033 82 fluid ounce ≡ 0.00026 U.S. fluid gallon ≡ 0.002 113 U.S. fluid pint

	≡ 0.271 fluid dram ≡ 16.231 minims (imperial)
1 cubic decimeter	≡ 61.024 cubic inches
1 kiloliter	≡ 1000 liters ≡ 10 hectoliters
1 dekaliter	≡ 10 liters ≡ 2.642 U.S. gallons ≡ 1.135 pecks
1 hectoliter	≡ 10 dekaliters ≡ 100 liters ≡ 26.418 U.S. gallons ≡ 2.838 bushels
1 centiliter	≡ 10 milliliters ≡ 0.01 liter ≡ 0.6103 cubic inch ≡ 2.705 drams ≡ 0.3382 U.S. fluid ounce
1 deciliter	≡ 10 centiliters ≡ 100 milliliters
1 milliliter (also cubic centimeter) ♫: 1 milliliter of pure water at 4 °C and one standard atmosphere pressure weighs exactly 1 gram	≡ 0.000 001 cubic meter ≡ 0.001 liter ≡ 0.202 884 135 U.S. teaspoon ≡ 0.004 226 752 82 U.S. cup ≡ 0.061 cubic inch ≡ 0.033 82 fluid ounce ≡ 0.000 26 U.S. fluid gallon ≡ 0.002 113 U.S. fluid pint ≡ 0.271 fluid dram ≡ 16.231 minims
1 minim (imperial)	≈ about 1 drop ≡ 1/480 fluid ounce (imperial) ≡ 1/60 fluid dram (imperial) ≡ 59.193 8802 × 10^{-9} cubic meter
1 minim (U.S.) (the smallest unit of fluid measure used in the United States of America)	≈ about 1 drop ≡ 1/480 fluid ounce (U.S.) ≡ 1/60 fluid dram (U.S.) ≡ 61.611 5199 × 10^{-9} cubic meter
1 lambda (λ)	≡ 1 cubic millimeter ≡ 1 × 10^{-9} cubic meter

1 metric cup	≡ 250 milliliters
1 drop (gtt) U.S.	≡ 1/360 U.S. fluid ounce ≡ 82.148 693 229 16 x 10^{-9} cubic meter
1 drop (metric)	≡ 1/20 milliliter ≡ 50.0 x 10^{-9} cubic meter
1 drop (medical)	≡ 1/12 milliliter ≡ 83.03 x 10^{-9} cubic meter
1 drop (gtt) (imperial)	≡ 1/288 fluid ounce (imperial) ≡ 98.656 467 01 x 10^{-9} cubic meter
1 fluid scruple (imperial)	≡ 1/24 fluid ounce (imperial) ≡ 1.183 877 60 x 10^{-6} cubic meter
1 fluid drachm (fl dr) (imperial)	≡ 1/8 fluid ounce (imperial) ≡ 3.551 632 81 x 10^{-6} cubic meter
1 fluid dram (fl dr) (U.S. fluidram)	≡ 1/8 fluid ounce ≡ 0.2256 cubic inch ≡ 3.696 691 195 3125 milliliters ≡ 60 minims ≡ 1.041 imperial fluid drachms
1 gill (gi U.S.) (U.S.)	≡ 4 fluid ounces (U.S.) ≡ 118.294 118 x 10^{-6} cubic meter
1 gill (gi imp) (imperial); noggin	≡ 5 fluid ounces (imperial) ≡ 142.065 312 x 10^{-6} cubic meter
1 teaspoon (tsp) (U.S. customary)	≡ 1/6 U.S. fluid ounce ≡ 1/3 tablespoon ≡ 1-1/3 fluid drams ≡ 4.928 921 595 milliliters
1 teaspoon (tsp) (imperial)	≡ 1/24 gill (imperial) ≡ 5.919 388 02083 milliliters
1 teaspoon (metric)	≡ 5.0 milliliters
1 teaspoon (U.S. food nutrition labeling	≡ 5.0 milliliters
1 tablespoon (tbsp) (U.S. customary)	≡ 3 teaspoons ≡ 1/2 U.S. fluid ounce ≡ 4 fluid drams ≡ 14.786 764 7825 milliliters
1 tablespoon (tbsp) (metric)	≡ 15.0 milliliters

1 tablespoon (tbsp) (imperial)	≡ 5/8 fluid ounce (imperial) ≡ 17.758 164 0625 milliliters
1 tablespoon (tbsp) (U.S. food nutrition labeling)	≡ 15 milliliters
3 tablespoons	≡ 1 jigger (bartending) ≡ 1.5 fluid ounces ≡ 44.360 milliliters
1 jigger (bartending)	≡ 1-1/2 fluid ounces (U.S.) ≡ 3 tablespoons ≡ 44.36 x 10^{-6} cubic meter
1 shot	≡ 1 U.S. fluid ounce ≡ 29.57 x 10^{-6} cubic meter
4 tablespoons	≡ 1 whiskey glass ≡ 2 fluid ounces ≡ 1/4 cup ≡ 59.147 milliliters
8 tablespoons	≡ 1 gill ≡ 4 fluid ounces ≡ 1/2 cup ≡ 1/4 pint ≡ 118.294 milliliters
1 pony	≡ 3/4 fluid ounce (U.S.) ≡ 22.180 147 171 875 milliliters
1 ounce (fl oz U.S.) (fluid U.S.)	≡ 2 tablespoons ≡ 0.0625 fluid pint ≡ 0.031 25 fluid quart ≡ 1/128 gallon (U.S.) ≡ 8 fluid U.S. drams ≡ 1.8047 cubic inches ≡ 1 pony ≡ 29.573 529 5625 milliliters ≡ 1.041 fluid ounces
1 ounce (fl oz U.S.) (fluid U.S. food nutrition labeling)	≡ 30 milliliters
1 pinch (fluid U.S.)	≡ 1/448 fluid ounce (U.S.) ≡ 616.115 199 x 10^{-9} cubic meter
1 ounce (fl oz imp) (fluid imperial)	≡ 1/160 gallon (imperial) ≡ 28.413 0625 milliliters

1 cup (c U.S.) (U.S. customary)	≡ 8 ounces (fluid) ≡ 16 tablespoons ≡ 1/2 pint ≡ 236.588 2365 milliliters ≡ 0.236 588 2365 liter
1 cup (c U.S.) U.S. food nutrition labeling	≡ 240 milliliters
1 cup (c) metric	≡ 250.0×10^{-6} cubic meter
1 pint (U.S.)	≡ 16 fluid ounces ≡ 1/8 gallon (U.S.) ≡ 2 cups ≡ 28.875 cubic inches ≡ 4 gills ≡ 128 fluid drams ≡ 32 tablespoons ≡ 0.47318 liter ≡ 473.176 475 milliliters
1 pint (pt imp) (imperial)	≡ 1/8 gallon (imperial) ≡ 568.261 25 milliliters
1 pottle; quarter	≡ 1/2 gallon (imperial) ≡ 80 fluid ounces (imperial) ≡ $2.273\ 045 \times 10^{-3}$ cubic meter
1 quart (qt U.S.) (U.S.)	≡ 1/4 gallon (fluid U.S.) ≡ 2 pints ≡ 32 fluid ounces ≡ 57.75 cubic inches ≡ 256 fluid drams ≡ 0.946 352 946 liter ≡ 946.352 946 milliliters ≡ 0.833 imperial quart
1 quart (qt imp) (imperial)	≡ 1.201 U.S. fluid quarts ≡ 1.1365 liters ≡ 69.354 cubic inches ≡ 1/4 gallon (imperial) ≡ $1.136\ 5225 \times 10^{-3}$ cubic meter
1 fifth	≡ 1/5 U.S. gallon ≡ $757.082\ 3568 \times 10^{-6}$ cubic meter
1 firkin	≡ 9 U.S. gallons ≡ 0.034 068 706 056 cubic meter

1 U.S. gallon	≡ 3.785 41 liters ≡ 3785.41 milliliters ≡ 0.8327 imperial gallon ≡ 231 cubic inches ≡ 0.1337 cubic foot ≡ 4 fluid quarts ≡ 8 pints ≡ 32 gills ≡ 128 fluid ounces ≡ 1024 fluid drams ≡ 8.3453 pounds of water ≡ 3.629 kilograms of water
1 gallon (gal imp) (imperial)	≡ 1.201 U.S. gallons ≡ 4.546 09 liters ≡ 277.42 cubic inches ≡ 160 imperial fluid ounces
1 gallon (beer gal) (beer)	≡ 282 cubic inches ≡ 4.621 152 04 x 10^{-3} cubic meter
1 gallon (gal US.) U.S. wine	≡ 231 cubic inches
1 butt, pipe	≡ 126 gallons wine ≡ 0.476 961 884 784 cubic meter
1 tun	≡ 252 gallons wine ≡ 0.953 923 769 568 cubic meter
1 kilderkin	≡ 18 gallons (imperial) ≡ 0.081 829 62 cubic meter
1 last	≡ 80 bushels (imperial) ≡ 2.909 4976 cubic meters
1 load	≡ 50 cubic feet ≡ 1.415 842 3296 cubic meters
1 imperial pint	≡ 1.201 U.S. fluid pints ≡ 34.6774 cubic inches ≡ 568.261 485 milliliters ≡ 20 imperial ounces ≡ 1/8 imperial gallon ≡ 1/2 imperial quart ≡ 4 imperial gills
U.S. gallons in square or rectangular tank	Multiply the number of cubic feet in the tank by 7.4805

U.S. gallons in circular tank	Square the diameter in feet, multiply by depth in feet, then multiply product by 5.875

VOLUME (DRY)

1 liter	≡ 0.9081 dry quart ≡ 1000 milliliters ≡ 0.1135 peck ≡ 0.02838 bushel
1 pinch (U.S.)	≡ 1/8 teaspoon (U.S.)
1 pinch (imperial)	≡ 1/8 teaspoon (imperial)
1 pint (pt U.S. dry) (dry U.S.)	≡ 33.600 cubic inches ≡ 1/64 bushel (U.S. level) ≡ 1/8 gallon (U.S. dry) ≡ 0.5 quart (U.S. dry) ≡ 0.551 liter ≡ 550.610 471 3575 milliliters
1 quart (qt U.S. dry) (dry U.S.)	≡ 1/32 bushel (U.S. level) ≡ 1/4 gallon (U.S. dry) ≡ 2 pints (U.S. dry) ≡ 67.2006 cubic inches ≡ 1.101 220 942 715 liters ≡ 0.969 imperial quart
1 quart (qt imp) imperial	≡ 1.0321 U.S. quarts ≡ 1.1365 liters ≡ 69.354 cubic inches
1 gallon (gal U.S.) (U.S. dry)	≡ 1/8 level U.S. bushel ≡ 4.404 883 77 x 10^{-3} cubic meter
1 peck (pk) (U.S. dry)	≡ 1/4 level bushel (U.S.) ≡ 16 pints (dry) ≡ 8 quarts (dry) ≡ 537.605 cubic inches ≡ 8.809 767 541 72 liters
1 bushel (bu imp) (imperial)	≡ 4 pecks ≡ 2150.42 cubic inches ≡ 1.2445 cubic feet ≡ 32 dry quarts

	≡ 64 pints
	≡ 35.238 liters
	≡ 0.036 368 72 cubic meter
1 bushel (bu U.S. lvl) U.S. dry level	≡ 2150 42 cubic inches
	≡ 0.035 239 166 88 cubic meter
1 bushel (bu U.S.) U.S. dry heaped	≡ 1-1/4 U.S. level bushels
	≡ 0.44 048 837 7086 cubic meter
1 strike (U.S.)	≡ 2 U.S. level bushels
	≡ 0.070 478 140 333 cubic meter
1 strike (imperial)	≡ 2 bushels (imperial)
	≡ 0.072 737 44 cubic meter
1 seam	≡ 8 U.S. level bushels
	≡ 0.281 912 561 331 cubic meter
1 coomb(e) (H) An old English measure of corn and wheat	= 4 bushels (imperial)
	= 1 quarter
	= 0.145 474 88 cubic meter
1 sack (U.S.)	≡ 3 U.S. level bushels
	≡ 0.105 717 210 500 cubic meter
1 sack (imperial); bag	≡ 3 bushels (imperial)
	≡ 0.109 106 16 cubic meter
1 wey (U.S.)	≡ 40 U.S. level bushels
	≡ 1.409 562 806 675 cubic meters
1 barrel (bl US) (U.S. dry)	≡ 105 quarts (dry) U.S.
	≡ 7056 cubic inches
	≡ 3.281 bushels
	≡ 115.628 198 985 liters
	≡ 0.115 628 198 985 cubic meter
1 register ton (a unit of volume used for the cargo capacity of a ship)	≡ 100 cubic feet
	≡ 2.831 684 6592 cubic meters
1 board-foot (fbm)	≡ 144 cubic inches
	≡ 2.359 737 216 x 10^{-3} cubic meter
1 timber foot	≡ 1 cubic foot
	≡ 0.028 316 846 592 cubic meter
1 cord-foot	= 16 cubic feet
	= 0.453 069 545 472 cubic meter
1 ton (freight)	≡ 40 cubic feet
	≡ 1.132 673 863 68 cubic meters

1 ton (displacement)	≡ 35 cubic feet ≡ 0.991 089 630 72 cubic meter
1 cord (firewood) (4 feet x 4 feet x 8 feet)	= 128 cubic feet = 3.624 556 363 776 cubic meters ♪: according to the <u>Farmer's Almanac</u>: there is enough space in a cord of firewood for a rat to squeeze through logs, but not for a cat to follow

DENSITY OF LIQUIDS, SOLIDS, AND SOIL

weight (pounds) of 1 U.S. gallon of anything x 7.5	≡ pounds per cubic foot
1 gallon water	≡ 8.3453 pounds ≡ 3.629 kilograms
1 quart water	≡ 2.086 pounds ≡ 0.907 25 kilogram
1 pint water	≡ 1.043 pounds ≡ 0.4536 kilogram
1 cubic-foot water	≡ 7.4805 gallons ≡ 62.428 pounds ≡ 28.32 kilograms
1 acre-foot water	≡ 2 719 886.4 pounds ≡ 1 233 808.47 kilograms
1 pound water	≡ 27.68 cubic inches water ≡ 0.1198 gallon
27.68 cubic inch water	≡ 1 pound
1 liter water	≡ 1 kilogram
1 milliliter water	≡ 1 gram
1 gram per milliliter (g/mL)	≡ 1000 kilograms per cubic meter
1 kilogram per liter (kg/L)	≡ 1000 kilograms per cubic meter
1 ounce per cubic foot (oz/ft^3)	≈ 1.001 153 961 kilograms per

(avoirdupois)	cubic meter
1 ounce per cubic inch (oz/in^3) (avoirdupois)	≈ 1.729 994 044 x 10^3 kilograms per cubic meter
1 ounce (avoirdupois) per gallon (oz/gal) (imperial)	≈ 6.236 023 291 kilograms per cubic meter
1 ounce (avoirdupois) per gallon (oz/gal) (U.S. fluid)	≈ 7.489 151 707 kilograms per cubic meter
1 pound (avoirdupois) per cubic foot (lb/ft^3)	≈ 16.018 463 37 kilograms per cubic meter
1 pound (avoirdupois) per cubic inch (lb/in^3)	≈ 2.767 990 471 x 10^4 kilograms per cubic meter
1 pound (avoirdupois) per gallon (lb/gal) (U.S. fluid)	≈ 119.826 4273 kilograms per cubic meter
1 pound (avoirdupois) per gallon (lb/gal) (imperial)	≈ 99.776 372 66 kilograms per cubic meter
1 slug per cubic foot (slug/ft^3)	≈ 515.378 8184 kilograms per cubic meter
1 acre-furrow-slice soil (top six inches)	≈ 2 million pounds (approximate) [when a soil has a common-found and typical agricultural/ horticultural dry bulk density of 1.47 g/cm^3] ♪: The dry bulk density of soil expresses the ratio of the mass of dried soil to its total volume (solids and pores together)
1 acre-foot soil (top 12 inches)	≈ 4 million pounds (approximate) [when a soil has a common-found and typical agricultural/ horticultural dry bulk density of 1.47 g/cm^3] ♪: The dry bulk density of soil expresses the ratio of the mass

	of dried soil to its total volume (solids and pores together)
1 hectare-furrow-slice soil (top 15 centimeters)	≈ 2 million kilograms (approximate)

TEMPERATURE

Degrees Fahrenheit from Celsius	°F ≡ 1.8 x (degrees Celsius) + 32
	Example: Convert 25 °Celsius to Fahrenheit: 1.8 x 25 + 32 ≡ 77 °Fahrenheit
Degrees Celsius from Fahrenheit	°C ≡ (Degrees Fahrenheit - 32) ÷ 1.8
	Example: Convert 77 °Fahrenheit to Celsius: (77-32) ÷ 1.8 ≡ 25 °Celsius
Degrees Celsius from kelvin	°C ≡ (kelvin − 273)
kelvin from Celsius	K ≡ (degrees Celsius) + 273
kelvin from Fahrenheit	K ≡ 5/9 (Fahrenheit degrees -32) + 273
Degrees Fahrenheit from kelvin	°F ≡ (kelvin degrees - 273) x (9/5) +32
K	≡ -273.15 °C
K	≡ -459.67 °F
normal human body temperature	≡ 98.6 °F

LUMINANCE; AND LUMINOUS INTENSITY AND FLUX

1 candela (cd) (SI base unit); candle	= 1 candela

The luminous intensity, in a given direction, of a source that emits monochromatic radiation of frequency 540 x 10^{12} hertz, and that has a radiant intensity in that direction of 1/683 watt per steradian	
candlepower (cp) (new) ♫: The use of candlepower as a unit is discouraged due to it ambiguity	= 1 candela
candlepower (cp) (old; pre-1948) ♫: Varies and is poorly reproducible	≈ 0.981 candela
1 candela per square foot (cd/ft^2)	≈ 10.763 910 417 candelas per square meter
1 candela per square inch (cd/in^2)	≈ 1550.0031 candelas per square meter
1 candela per square meter (cd/m^2) (SI base unit)	= 1 candela per square meter
1 foot lambert (fL)	≡ (1/π) candela per square foot ≈ 3.426 259 0996 candelas per square meter
1 lambert (L)	≡ (10^4/π) candelas/square meter ≈ 3183.098 8618 candelas per square meter
1 stilb (sb) (CGS unit)	≡ 10^4 candelas per square meter ≈ 1 x 10^4 candelas per square meter
1 lumen (lm) (SI base unit)	≡ candela·solid angle = 1 lumen
1 footcandle; lumen per square foot (fc)	≡ 1 lumen per square foot = 10.763 910 417 lux
1 lumen per square inch (lm/in^2)	≡ lumen per square inch/ ≈ 1 550.0031 lux
1 lux (SI base unit) (lx)	≡ lumen per square meter = 1 lux
1 phot (ph) (CGS unit)	≡ lumen/mc^2 = 1 x 10^4 lux

WORK, POWER, AND ENERGY; OR HEAT WORK

♪: for below conversions: any unit of <u>energy</u> divided by any unit of <u>time</u> is a unit of <u>power</u>	
1 joule (J) (SI base unit)	= The work done when a force of one newton moves the point of its application a distance of one meter in the direction of the force. ≡ 0.738 foot pound ≡ 107 ergs ≡ 0.239 calorie
1 Celsius heat unit (International Table (CHU$_{IT}$)	= 1 BTU$_{IT}$ × 1 K/ °R = 1.899 100 534 716 × 10^3 joules
1 calorie (cal$_{IT}$) (International Table)	≡ 4.1868 joules ≡ 0.003 968 British thermal unit (BTU$_{IT}$)
1 calorie (cal$_{mean}$) (mean)	= 1/100 of the energy required to warm one gram of air-free water from 0 °C to 100 °C @ 1 atmosphere ≈ 4.190 02 joules
1 calorie (cal$_{th}$) (thermochemical)	≡ 4.184 joules
1 calorie (cal$_{3.98 °C}$) 3.98 °C	≈ 4.2045 joules
1 calorie (cal$_{15 °C}$) 15 °C	≈ 4.1855 joules
1 calorie (cal$_{20 °C}$) 20 °C	≈ 4.1819 joules
1 kilocalorie; large calorie (kcal; Cal)	≡ 1000 cal$_{IT}$ = 4.1868 × 10^{-3} joule
1 calorie (cal$_{IT}$/s) (International Table) per second	≡ 1 (cal$_{IT}$/s) = 4.1868 watts
1 therm (E.C.)	≡ 100 000 BTU$_{IT}$ = 105.505 585 262 × 10^6 joules
1 therm (U.S.)	≡ 100 000 BTU$_{59 °F}$ = 105.4804 × 10^6 joules

1 thermie (th)	$\equiv$ 1 Mcal$_{IT}$ = 4.1868 x 10^6 joules
1 British Thermal Unit (BTU$_{IT}$) (International Table)	$\equiv$ 778.16 foot pounds $\equiv$ 252 calories $\equiv$ 1055 joules $\equiv$ 0.252 kilogram calories $\equiv$ 10.409 liter atmospheres $\equiv$ 1.055 x 10^{10} ergs $\equiv$ 0.0003927 horsepower hour
1 British Thermal Unit (BTU$_{ISO}$) (ISO)	= 1.0545 x 10^3 joules
1 British Thermal Unit (BTU$_{mean}$) (mean)	$\approx$ 1.055 87 x 10^3 joules
1 British Thermal Unit (BTU$_{th}$) (thermochemical)	$\approx$ 1.054 350 x 10^3 joules
1 British Thermal Unit (BTU$_{39 °F}$) 39 °F	$\approx$ 1.059 67 x 10^3 joules
1 British Thermal Unit (BTU$_{59 °F}$) 59 °F	$\approx$ 1.054 804 x 10^3 joules
1 British Thermal Unit (BTU$_{60 °F}$) 60 °F	$\approx$ 1.054 68 x 10^3 joules
1 British Thermal Unit (BTU$_{63 °F}$) 63 °F	$\approx$ 1.0546 x 10^3 joules
1 British Thermal Unit (BTU$_{IT}$) per hour	$\equiv$ 1 BTU$_{IT}$/h $\approx$ 0.293 071 watt $\equiv$ 0.07 gram calorie per second $\equiv$ 0.2162 foot pound per second $\equiv$ 0.00039 horsepower
1 British Thermal Unit (BTU$_{IT}$) per minute	$\equiv$ 1 BTU$_{IT}$/min $\approx$ 17.584 264 watts $\approx$ 0.017 584 264 kilowatt $\equiv$ 0.023 56 horsepower $\equiv$ 12.96 foot pounds per second
1 British Thermal Unit (BTU$_{IT}$) per second	$\equiv$ 1 BTU$_{IT}$/s $\approx$ 1.055 055 852 62 x 10^3 watts

1 horsepower (hp) (imperial electrical) ♪: the usage of CC for cm³ is incorrect	≡ 0.746 kilowatt (kW) ≡ 746 watts ≡ 550 foot pounds per second ≡ 2544.43 British Thermal Units (BTU_{IT}) per hour Also: there is 1 horsepower for every 15-17 cubic centimeters (cm³) in an engine (approximate)
1 horsepower (bhp) (boiler)	≈ 34.5 lb/h x 970.3 BTU_{IT}/lb ≈ 9.810 657 x 10³ watts
1 horsepower (hp) (European electrical)	= 75 kp·m/s ≈ 736 watts
1 horsepower (hp) (imperial mechanical)	≡ 550 ft lbf/s = 745.699 871 270 22 watts
1 horsepower (hp) (metric)	≡ 75 m ft kgf/s = 735.498 75 watts
1 horsepower-hour (hp·h)	= 2.684 519 537 696 x 10⁶ joules
1 barrel of oil equivalent (bboe)	≈ 5.8 x 10⁶ $BTU_{59 °F}$ ≈ 6.12 x 10⁹ joule
1 ton of oil equivalent (TOE)	≈ 10 $Gcal_{TH}$ ≈ 6.12 x 10⁹ joule
1 ton of coal equivalent (TCE)	≡ 7 $Gcal_{th}$ = 41.868 x 10⁹ joules
1 ton of TNT (tTNT)	≡ 1 $Gcal_{th}$ = 4.184 x 10⁹ joules
1 cubic foot of natural gas	≡ 1000 (BTU_{IT}) = 1.055 055 852 62 x 10⁶ joules
1 kilowatt	≡ 3,415 British Thermal Unit (BTU_{IT}) per hour
1 watt (W) (Si base unit) (a unit of power)	= The power which in one second of time gives rise to one joule of energy ≡ 1 volt x 1 ampere ≡ 0.738 foot pounds per second ≡ 3.415 British Thermal Unit

	(BTU$_{IT}$) per hour ≡ 0.00134 horse power ≡ one joule per second (joule is the unit of energy) = 1 newton·meter per second = 1 kg·m^2/s^3
1 kilowatt (a unit of power)	≡ 1.34 horse power ≡ 1000 watts
1 kilowatt-hour; (kW·h)(Board of Trade Unit)	≡ 1 kW x 1 hour = 3.6 x 10^6 joules ≡ 1000 watt-hours
1 inch-pound force (in lbf)	≡ g x 1 lb x 1 in ≡ 0.112 984 829 027 6167 joule
1 foot-pound force (ft lbf)	≡ g x 1 lb x 1 ft ≡ 1.355 817 948 331 4004 joules
1 foot-pound force per hour (ft lbf/h)	≡ 1 ft lbf/h = 3.766 161 x 10^{-4} watt
1 foot-pound force per minute (ft lbf/min)	≡ 1 ft lbf/min = 2.259 696 580 552 x 10^{-4} watt
1 foot-pound force per second (ft lbf/sec)	≡ 1 ft lbf/sec = 1.355 817 948 331 4004 watts
1 foot-poundal (ft pdl)	≡ 1 lb·ft^2/s^2 = 4.214 011 009 380 48 x 10^{-2} joule
1 electronvolt (eV)	≡ e x 1 volt ≡ 1.602 177 33 x 10^{-19} joule
1 erg (CGS unit) (centimeter–gram–second system)	≡ 1 g·cm^2/s^2 = 10^{-7} joule
1 erg per second (erg/s)	≡ 10^{-7} watt
1 lusec (lusec)	≡ 1 L· µmHg/s ≈ 1.333 x 10^{-4} watt
1 poncelet (p)	≡ 100 m kgf/s = 980.665 watts
1 square foot equivalent direct radiation (sq ft EDR)	≡ 240 BTU$_{IT}$/h = 70.337 057 watts

1 ton of air conditioning	≡ 1 ton ice melted/24 hours ≈ 3504 watts
1 ton of refrigeration (imperial)	≡ 1 BTU$_{IT}$ x 1 ng tn/lb ÷ 10 min/s ≈ 3.938 875 x 10^3 watts
1 ton of refrigeration (IT)	≡ 1 BTU$_{IT}$ x 1 sh tn/lb ÷ 10 min/s ≈ 3.516 853 x 10^3 watts
1 liter of atmosphere; (1 atm; sl)	≡ 1 atmosphere x 1 cubic centimeter = 0.101 325 joule
1 liter-atmosphere (l·atm/min) per minute	= 1.688 75 watts
1 liter-atmosphere (l·atm/sec) per second	= 101.325 watts
1 cubic centimeter of atmosphere; standard cubic centimeter (cc atm; scc)	≡ 1 atmosphere x 1 liter = 101.325 joules
1 atmosphere-cubic centimeter per minute (atm ccm)	≡ 1 atm x 1 cm^3/min = 1.688 75 x 10^{-3} watt
1 atmosphere-cubic centimeter per second (atm ccs)	≡ 1 atm x 1 cm^3/s = 0.101 325 watt
1 atmosphere-cubic foot per hour (atm cfh)	≡ 1 atm x 1 cu ft/h = 0.797 001 244 704 watt
1 atmosphere-cubic foot per minute (atm cfmin)	≡ 1 atm x 1 cu ft/min = 47.820 074 682 24 watts
1 atmosphere-cubic foot per second (atm cfs)	≡ 1 atm x 1 cu ft/sec = 2.869 204 480 934 x 10^3 watts
1 gallon-atmosphere (U.S. gal atm) (U.S.)	≡ 1 atmosphere x 1 gallon (U.S.) = 383.556 849 0138 joules
1 gallon-atmosphere (imp gal atm) (imperial)	≡ 1 atmosphere x 1 gallon (imperial) = 460.632 569 25 joules
1 cubic foot of atmosphere; standard cubic foot (cu ft atm; scf)	≡ 1 atmosphere x 1 cubic foot = 2.869 204 480 9344 x 10^3 joules
1 cubic yard of atmosphere; standard	≡ 1 atmosphere x 1 cubic yard

cubic yard (cu yd atm; scy)	= 77.468 520 985 22 x 10^3 joules
1 quad	≡ 10^{15} BTU_{IT} = 1.055 055 852 62 x 10^{18} joules
1 hartree, atomic unit of energy (E_h)	= 4.359 744 x 10^{-18} joule
1 rydberg (Ry)	≈ 2.179 872 x 10^{-18} joule

DYNAMIC AND KINEMATIC VISCOSITY

1 pascal second (Pa·s) (SI base unit)	≡ N·s/m^2, kg/(m·s) = 1 Pa·s
1 poise (P) (cgs unit) (centimeter–gram–second system)	≡ 1 barye·s = 0.1 Pa·s
1 pound per foot hour (lb/(ft·h))	≡ 1 lb/(ft·h) ≈ 4.133 789 x 10^{-4} Pa·s
1 pound per foot second (lb/(ft·s))	≡ 1 lb/(ft·s) ≈ 1.488 164 Pa·s
1 pound-force second per square foot (lbf·s/ft^2)	≡ 1 lbf·s/ft^2 ≈ 47.880 26 Pa·s
1 pound-force second per square inch (lbf·s/in^2)	≡ 1 lbf·s/in^2 ≈ 6 894.757 Pa·s
square foot per second (ft^2/s)	≡ 1 ft^2/s = 0.092 903 04 m^2/s
square meter per second (m^2/s) (SI base unit)	≡ 1 m^2/s
1 stokes (St) (cgs unit) (centimeter–gram–second system)	≡ 10^{-4} m^2/s

PRESSURE

1 pascal (Pa) (SI base unit)	≡ 0.000 010 197 162 13 kilogram per square meter ≡ 1 newton per square meter ≡ 1.45 x 10^{-4} pound per square inch

1 pound-force per square inch (psi)	≡ 1 lbf/in² ≡ 6.894 757 293 x 10³ pascals ≡ 6.894 757 293 kilopascals ≡ 0.0703 kilogram per square centimeter ≡ 0.06895 bar ≡ 0.068 045 957 0643 atmosphere ≡ 2.036 inches mercury
1 pound-force per square foot (psf)	≡ 1 lbf/ft² = 47.880 26 pascals
1 ton (short) per square foot	≡ 1 short ton x *g* / 1 square foot ≈ 95.760 518 x 10³ pascals
1 poudal per square foot (pdl/sqft)	= 1.488 164 pascals
1 kilogram-force per square millimeter (kfg/mm²)	≡ 9.806 65 x 10⁶ pascals
1 kilopascal (kPa) (Si unit)	≡ 0.145 037 891 149 pound-force per square inch
1 bar (bar)	≡ 10⁵ newtons per square meter ≡ 1000 millibars ≡ 10 000 (10⁵) pascals ≡ 0.9869 atmosphere ≡ 29.53 inches of mercury ≡ 14.5 pounds-force per square inch ≡ 1.0 x 10⁶ dynes per square centimeter ≡ 10 200 kilograms per square meters ≡ 2089 pounds-force per square foot
1 barye	≡ 1 dyne per square centimeter ≡ 0.1 pascal
1 torr (torr)	≡ 101 325/750 pascals ≈ 133.3224 pascals
1 kip per square inch (ksi)	≡ 1 kipf/sq in ≡ 6.894 757 x 10⁶ pascals
1 millibar	≡ 0.001 bar ≡ 100 pascals ≡ 0.02953 inch of mercury

	≡ 0.000 986 9 atmosphere
1 micron (micrometer) of mercury (µmHg)	≡ 13 595.1 kg/m^3 x 1 µm x g ≈ 0.001 torr ≈ 0.133 3224 pascal
1 millimeter of mercury (mmHg)	≡ 13 595.1 kg/m^3 x 1mm x g ≈ 1 torr
1 centimeter of mercury (cmHg)	≡ 13 595.1 kg/m^3 x 1 cm x g ≈ 1.333 22 x 10^3 pascals
1 inch of mercury (inHg) conventional)	≡ 13 595.1 kg/m^3 x 1in x g ≡ 0.033 86 bar ≡ 33.86 millibars ≈ 3 386 389 pascals ≡ 0.033 42 atmosphere
1 foot of mercury (ftHg) (conventional)	≡ 13 595.1 kg/m^3 x 1 ft x g ≈ 40.636 66 x 10^3 pascals
1 centimeter of water (4 °C)	≈ 999.972 kg/m^3 x 1 cm x g ≈ 98.0638 pascals
1 millimeter of water (mmH$_2$O) (3.98 °C)	≈ 999.972 kg/m^3 x 1mm x g = 0.999 972 kgf/m^2
1 inch of water (inH$_2$O) (39.2 °F)	≈ 999.972 kg/m^3 x 1 in x g ≈ 249.082 pascals
1 foot of water (ftH$_2$O) (39.2 °F)	≈ 999.972 kg/m^3 x 1 ft x g ≈ 2.988 98 x 10^3 pascals
1 atmosphere (atm) (standard)	≡ 29.92 inches of mercury ≡ 101 325 pascals (or newtons) ≡ 101.325 kilopascals (kPa) ≡ 1013 millibars ≡ 1.013 bars ≡ 760 torr ≡ 14.70 pounds-force per square inch ≡ 0.007 348 ton per square inch ≡ 1.058 tons per square foot ≡ 1.0333 kilograms per square centimeter ≡ 760 millimeters mercury (at 0 °C) ≡ 76.0 centimeters of mercury (at 0 °C)

	≡ 33.9 feet of water (at 4 °C)
1 atmosphere (at) (technical)	≡ 1 kilogram-force per square centimeter (kgf/cm^2) ≡ 9.806 65 x 10^4 pascals
1 pièze (pz) (a meter-tonne-second [mts] unit)	≡ 1000 kg/m·s^2 = 1 x 10^3 pascals = 1 kilopascal
air pressure at sea level is approximately	≡ 1 atmosphere ≡ 1 bar ≡ 105 newtons per square meter (or pascals) ≡ 14.50 pounds per square inch
pressure (in pounds per square inch)	≡ height (feet) x 0.434

TORQUE OR MOMENT OF FORCE

foot-pound force (ft lbf)	≡ g x 1 lb x 1 ft = 1.355 817 948 331 4004 newtons·meter
foot-poundal (ft pdl)	≡ 1 lb·ft^2/s^2 = 4.214 011 009 380 48 x 10^{-2} newton·meter
inch-pound force (in lbf)	≡ g x 1 lb x 1 in = 0.112 984 829 027 6167 newton·meter
meter kilogram (m kg)	≡ N x m / g ≈ 0.101 971 621 newton·meter
newton meter (N) (SI base unit)	≡ N x m ≡ kg·m^2/s^2 = 1 newton·meter

VELOCITY AND SPEED (distance ÷ time)

1 kilometer per hour (km/h)	≡ 0.6214 mile per hour ≡ 0.912 foot per second ≡ 0.277 7778 meter per second
1 knot (kn,kt) (international)	≡ 1 nautical mile per hour ≡ 1.852 kilometers per hour ≈ 1.151 miles per hour

1 knot (kn) (Admiralty)	≡ 0.514 444 meter per second ≡ 1 nautical mile (Admiralty) per hour ≡ 1.853 184 kilometers per hour ≡ 0.514 773 meter per second
1 mile per hour (mph)	≡ 1.609 kilometers per hour ≡ 0.447 04 meter per second ≡ 44.70 centimeters per second ≡ 88 feet per minute ≡ 1.467 feet per second
1 mile per minute (mpm)	≡ 60 miles per hour ≡ 26.8224 meters per second
1 mile per second (mps)	= 1 609.344 meters per second
1 foot per hour (fph)	≡ 8.466 667 x 10^{-5} meter per second
1 inch per hour (iph)	≡ 7.05 556 x 10^{-6} meter per second
1 foot per minute (fpm)	≡ 0.011 36 mile per hour ≡ 5.08 x 10^{-3} meter per second
1 inch per minute (ipm)	≡ 4.23 333 x 10^{-4} meter per second
1 foot per second (fps)	≡ 0.3048 meter per second
1 inch per second (ips)	≡ 2.54 x 10^{-2} meter per second
1 meter per second (m/s)	≡ 6000 centimeters per minute ≡ 3.28 feet per second ≡ 2.237 miles per hour
1 centimeter per second	≡ 0.022 37 mile per hour ≡ 0.032 81 foot second ≡ 0.036 kilometer hour ≡ 0.019 43 knot
1 furlong per fortnight	≡ 1.663 095 x 10^{-4} meter per second
mach number (*M*). Ratio of the speed : the speed of sound in the medium. Varies especially with temperature.	≈ 1225 kilometers per hour in air at sea level; 1062 kilometers per hour at jet altitudes ≈ 761 miles per hour in air at sea level; 660 miles per hour at jet altitudes

	≈ 340 to 295 meters per second for aircraft
velocity of sound in dry air (at 0° Celsius and 1 atmosphere, and at sea level)	≡ 1087 feet per second ≡ 761 miles per hour ≡ 33 136 centimeters per second ≡ 331.36 meters per second
velocity of light (in a vacuum)	≡ 186 282.4 miles per second ≡ 299 792 458 meters per second ≡ 299 792.5 kilometers per second

<u>Speed of Thunder</u>: Thunder travels through air at the speed of sound (officially 1,087 feet per second in dry air at 32 °F [0 °C]). At normal summer temperatures, e.g. 80 °F, the speed is about 1200 feet per second, but also changes depending on humidity. At this temperature, sound travels one mile in roughly 5 seconds. Therefore, if lightning occurs one mile away, it takes thunder approximately five seconds to travel the mile, and for two miles it takes approximately ten seconds.

ACCERLERATION (velocity ÷ time)

1 meter per second squared (m/s^2) (SI base unit)	≡ 100 centimeters per second squared ≡ 3.280 84 feet per second squared ≡ 39.370 08 inch per second squared ≡ 0.000 621 371 2 mile per second squared ≡ 0.001 kilometer per second squared ≡ 1000 millimeters per second squared ≡ 100 Galileo
1 mile per hour per second (mph/s)	≡ 4.4704 x 10^{-1} meter per second squared
1 mile per minute per second (mpm/s)	≡ 26.8224 meters per second squared
1 mile per second squared (mps^2)	= 1.609 344 x 10^3 meters per second squared
1 foot per second squared (fps^2)	≡ 0.3048 meter per second squared

	≡ 30.5 centimeters per second squared
1 foot per hour per second (fph/s)	≈ 8.466 667 x 10⁻⁵ meter per second squared
1 foot per minute per second (fpm/s)	= 5.08 x 10⁻³ meter per second squared
1 inch per minute per second (ipm/s)	≈ 4.233 333 x 10⁻⁴ meter per second squared
1 inch per second squared (ips²)	≡ 2.54 x 10⁻² meter per second squared
1 Galileo; gal (Gal)	= 10⁻² meter per second squared
standard gravity (g)	≡ 9.806 65 meters per second squared
acceleration due to gravity on Earth	≡ 9.81 meters per second ≡ 981 centimeters per second ≡ 32.2 feet per second

FORCE (mass x acceleration)

1 newton (N) (SI base unit) A force capable of giving a mass of one kilogram, an acceleration of one meter per second, per second (or one meter per second squared)	≡ 1 kilogram per meter per second squared ≡ 0.224 808 943 871 pound
1 dyne (dyn)	= 10⁻⁵ newton
1 ton-force (tnf)	≡ 8.896 443 230 521 x 10³ newtons
1 kilogram-force; kilopond; grave-force (kgf; kp; Gf)	≡ 9.806 65 newtons
1 milligrave-force; gravet-force (mGf; gf)	≡ 9.806 65 millinewtons
1 pound-force (ozf)	≡ 16 ounces ≡ 4.448 221 615 2605 newtons
1 ounce-force	≡ 0.278 013 850 953 7812 newton
1 kip-force; kip (kip; kipf; klbf)	≡ 4.448 221 615 2605 x 10³ newtons

1 poundal (pdl)	= 0.138 254 954 376 newton
1 sthene (sn)	= 1 x 10³ newtons
1 atomic unit of force	≈ 8.238 722 06 x 10⁻⁸ newton

FREQUENCY; ANGULAR FREQUENCY; AND ANGULAR VELOCITY

Name of Unit	Definition	Relation to SI Units
hertz (Hz) (SI base unit)	≡ the number of cycles per second	= 1 hertz = 1 per second
revolutions per minute (rpm)	≡ one unit rpm equals one rotation completed around a fixed axis in one minute of time	≈ 0.104 719 755 radian per second
2 π radians per second		≡ 1 hertz (Hz)
1 radian per second (rad/s)	The SI unit of angular velocity. The radian per second is also the unit of angular frequency The change in the orientation of an object, in radians, every second	≈ 0.159155 Hz
1 radian per second		≈ 57.29578 degrees per second
1 radian per second		≈ 9.5493 revolutions per minute

ELECTRIC CURRENT AND CHARGE

1 ampere (A) (SI base unit). The constant current needed to produce a force of 2 x 10⁻⁷	≡ 1 A

newton per meter between two straight parallel conductors of infinite length and negligible circular cross-section placed one meter apart in a vacuum	
1 electronmagnetic unit; abampers (abamp) (cgs unit) (centimeter–gram–second system)	≡ 10 amperes
1 esu per second; statampers (esu/s) (cgs unit) (centimeter–gram–second system)	≡ (0.1 A·m/s)/c ≈ 3.335 641 x 10^{-10} ampere
1 coulomb (C) (SI base unit) The amount of electricity carried in one second of time by one ampere of current.	= 1 coulomb = 1 ampere·second
1 abcoulomb; electromagnetic unit (abC; emu) (cgs unit)	≡ 10 coulombs
1 atomic unit of charge (au)	≡ e ≈ 1.602 176 462 x 10^{-19} coulomb
1 faraday (F)	≡ 1 mol x N_A·e ≈ 96 485.3383 coulombs
1 statcoulomb; franklin; electrostatic unit (cgs unit) (statC; Fr; ese)	≡ (0.1 A·m)/c ≈ 3.335 641 x 10^{-10} coulomb

SI (metric) MULTIPLES FOR HERTZ (Hz)

Submultiples			Multiples		
Value	Symbol	Name	Value	Symbol	Name
10^{-1} Hz	dHz	decihertz	10^{1} Hz	daHz	decahertz
10^{-2} Hz	cHz	centihertz	10^{2} Hz	hHz	hectohertz
10^{-3} Hz	mHz	millihertz	10^{3} Hz	kHz	kilohertz
10^{-6} Hz	µHz	microhertz	10^{6} Hz	MHz	megahertz
10^{-9} Hz	nHz	nanohertz	10^{9} Hz	GHz	gigahertz
10^{-12} Hz	pHz	picohertz	10^{12} Hz	THz	terahertz
10^{-15} Hz	fHz	femtohertz	10^{15} Hz	PHz	petahertz
10^{-18} Hz	aHz	attohertz	10^{18} Hz	EHz	exahertz
10^{-21} Hz	zHz	zeptohertz	10^{21} Hz	ZHz	zettahertz
10^{-24} Hz	yHz	yoctohertz	10^{24} Hz	YHz	yottahertz

PLANE ANGLES

Name of Unit	Definition	Relation to SI Units
angular mil (µ)	$\equiv 2\pi/6400$ radians	$\approx 0.981\ 748 \times 10^{-3}$ radian
arcminute; MOA (')	$\equiv 1°/60$	$\approx 0.290\ 888 \times 10^{-3}$ radian
arcsecond (")	$\equiv 1°/3600$	$\approx 4.848\ 137 \times 10^{-3}$ radian
centesimal minute of arc (')	$\equiv 1$ gradian/100	$\approx 0.157\ 080 \times 10^{-3}$ radian
centesimal second of arc (")	$\equiv 1$ gradian/10 000	$\approx 1.570\ 796 \times 10^{-6}$ radian
degree of arc (°)	$\equiv 1/360$ of a revolution $\equiv \pi/180$ radians	$\approx 17.453\ 293 \times 10^{-3}$ radian
gradian; grad; gon (grad)	$\equiv 1/400$ of a revolution $\equiv 2\pi/400$ radians $\equiv 0.9°$	$\approx 15.707\ 963 \times 10^{-3}$ radian
octant	$\equiv 45°$	$\approx 0.785\ 398$ radian
quadrant	$\equiv 90°$	$\approx 1.570\ 796$
radian (rad) (SI base unit)	$\equiv$ The angle subtended at the centre of a circle by an arc equal in length to the radius of the circle.	1 radian $\equiv 57.3°$
sextant	$\equiv 60°$	$\approx 1.047\ 198$ radians
sign	$\equiv 30°$	$\approx 0.523\ 599$ radian

TIME (A QUICK REFERENCE)

Unit	Size	Notes
yoctosecond	10^{-24} second	
zeptosecond	10^{-21} second	
attosecond	10^{-18} second	shortest time now

		measurable
femtosecond	10^{-15} second	pulse time of ultrafast lasers
picosecond	10^{-12} second	
nanosecond	10^{-9} second	time for molecules to fluoresce
microsecond	10^{-6} second	
millisecond	0.001 second	
second	1 second	SI base unit
minute	60 seconds	
hour	60 minutes	
day	24 hours	
week	7 days	also called: a sennight
fortnight	14 days	2 weeks
lunar month	27.2–29.5 days	various definitions of lunar month exist
month	28–31 days	
quarter	3 months	
year	12 months	
common year	365 days	52 weeks + 1 day
leap year	366 days	52 weeks + 2 days
tropical year	365.242 19 days	average
Gregorian year	365.2425 days	average
Olympiad	4 year cycle	

lustrum	5 years	also called: a pentad
decade	10 years	
indiction	15 year cycle	
generation	17–35 years	approximate
Jubilee (Biblical. see Lev. 25:8-17)	50 years	
century	100 years	
millennium	1000 years	
exasecond	10^{18} seconds	roughly 32 billion years, more than twice the age of the universe on current estimates
cosmological decade	varies	10 x the length of the previous cosmological decade, with CÐ 1 beginning either 10 seconds or 10 years after the Big Bang, depending definition

MORE TIME DEFINITIONS AND CONVERSIONS

1 atomic unit (au) of time	$\approx 2.418\,884\,254 \times 10^{-17}$ second
1 Callippic cycle: a lunar cycle noted by Greek astronomer Callippus in 325 BC	= a period of four Metonic cycles = 76 Julian years = 940 months = 27 759 days = $2.398\,3776 \times 10^{9}$ seconds
1 century (c)	$\equiv$ 100 years (see definition for length of year)
1 day (d)	$\equiv$ 24 hours $\equiv$ 1400 minutes

1 day (d) (sidereal)	≡ 86 400 seconds ≡ the time needed for the Earth to rotate once on its axis ≡ 86 164.1 seconds
1 decade (dec)	≡ 10 years (see definition for length of a year)
1 fortnight (fn)	≡ 2 weeks ≡ 336 hours ≡ 20 160 minutes ≡ 1 209 600 seconds
1 helek (Hebrew)	≡ 1/1080 hour ≡ 3.3 seconds
1 Hipparchic cycle (a lunar cycle noticed by the Greek astronomer Hipparchus [190 BC – c. 120 BC])	≡ 4 Callippic cycles less one day ≡ 3759.9998 months ≡ $9.593\,424 \times 10^9$ seconds
1 hour (h)	≡ 60 minutes ≡ 3600 seconds
1 jiffy (j)	≡ 1/60 second ≡ 0.016 second
1 ke (quarter of an hour)	≡ 1/4 hour ≡ 15 minutes
1 lustrum; luster (H) (A ceremonial purification of the entire ancient Roman population after the census every five years)	= 5 years (of 365 day years) = 1.5768×10^8 seconds
1 Metonic cycle (or Enneadecaeteris) (from Greek words for nineteen years)	≡ 6940 days ≈ 19 years = 5.996×10^8 seconds
1 millennium	≡ 1000 years (see definition for length of a year)
1 milliday (md)	≡ 1/1000 day = $(24 \times 60 \times 60) \div 1000 = 86.4$ seconds
1 minute	≡ 60 seconds (however, due to leap seconds, 1 minute may be 59 seconds or 61 seconds)

1 moment	≡ 90 seconds
1 month (full) (mo)	≡ 30 days = 2 592 000 seconds
1 month (hollow) (mo)	≡ 29 days
octaeteris	48 full months + 48 hollow months + 3 full months = 8 years of 365.25 days = 2922 days = $2.524\,608 \times 10^8$ seconds
1 month (Greg.av.) (mo)	Average Gregorian month = 365.2425 days ÷ 12 months = 30.436 875 days = 2.6297×10^6 seconds
Planck time (P) [In physics, the time required for light to travel in a vacuum a distance of 1 Planck length: $1.616\,199 \times 10^{-35}$ meters]	≈ $1.351\,211\,868 \times 10^{-43}$ second
1 second (s) (SI base unit)	The International Systems of Measurement (SI) defines the second as: "9 192 631 770 cycles of that radiation which corresponds to the transition between two electron spin energy levels of the ground state of the ^{133}Cs (caesium) atom"
1 shake	≡ 10^{-8} second ≡ 10 nanoseconds
1 sigma	≡ 10^{-6} second ≡ 1 μ (micro) second
1 Svedberg unit (S; Sv)	≡ 10^{-13} second (a unit of sedimentation coefficient)
1 week (wk)	≡ 7 days = 168 hours = 10 080 minutes = 604 800 seconds
1 year (yr, y, or a) (Gregorian) ♪: Used throughout the entire world today. Also called	≡ 365.2425 days average, calculated from common years (365 days) plus leap years (366 days) on most years that are divisible by 4. ≡ 31 556 952 seconds

the Christian calendar	
1 year (yr, y, or a) (sidereal)	≡ The time taken by the Earth to orbit the Sun once with respect to the fixed stars. Therefore, it is also the time taken for the Sun to return to the same position with respect to the fixed stars after apparently travelling once around the ecliptic. = 365.256 363 days = 31 558 149.7632 seconds
1 year (yr, y, or a) (tropical)	≡ The length of time it takes for the Sun to return to the same position in the cycle of seasons. = 365.242 190 days = 31 556 925 seconds
1 year (a) Julian	= In astronomy, a Julian year (symbol: a) is a unit of measurement of time defined as exactly 365.25 days of 86 400 SI seconds each, totaling 31 557 600 seconds

PARTS PER MILLION (ppm)

Parts per million (ppm) is generally a measure of concentration which is used where low levels of concentrations are significant. The ppm value is equivalent to the absolute fractional amount multiplied by one million (10^6).

One part per million denotes one part per 1 000 000 parts, one part in 10^6, and a value of 1×10^{-6}. This is equivalent to one drop of water diluted into 50 liters (which is more or less the fuel tank capacity of a compact car), or about 32 seconds out of a year.

♪: Although the International Bureau of Weights and Measures (the international standards organization) recognizes the use of parts-per notation, it is not formally part of the International System of Units (SI).

1%	≡ 10 000 ppm
ppm ÷ by 10 000	≡ % (percent, or per centum)
percent x 10 000	≡ parts per million
ppm ÷ 1000 x equivalent weight	≡ milliequivalents per gram

ppm ÷ 10 equivalent weights	≡ milliequivalents per 100 grams
milliequivalents per liter x equivalent weight	≡ ppm
1 ppm	≡ 1 microgram per milliliter water
1 ppm ♪: 1 liter of pure water at 4 °C and one standard atmosphere pressure weighs exactly 1 kilogram, so 1 milligram/liter is ≡ 1 ppm	≡ 1 milligram per liter of solution (a concentration term). ♪: one part per million of a solid in a liquid can be written as a milligram per liter and abbreviated mg/l.
1 ppm	≡ 1 gram per 1000 liters
	≡ 1 one milliliter (or cubic centimeter) in a cubic meter (or kiloliter)
1 ppm	≡ 1 milligram per kilogram (mg/kg) (a mass term)
1 ppm	≡ 1 gram per megagram (or metric tonne)
1 ppm	≡ 1 microgram per gram
1 ppm	≡ 1 micrometer per meter
1 ppm	≡ 1 micromole per mole
1 ppm	≡ 8.345 pounds per million gallons water
1 ppm	≡ 0.379 gram per 100 U.S. gallons
1 ppm	≡ 0.013 ounce in 100 U.S. gallons water
1 ppm	≡ 1 gram per 1000 liters
0.95 ppm	≈ one square centimeter in 1000 square feet

Four drops of ink in a 55 gallon (208 liters) barrel of water would produce an "ink concentration" of ≈ 1 ppm.

RADIOACTIVITY

1 curie	≡ 3.70 x 10^{10} disintegration per second
	≡ 3.70 x 10^{10} bequerels

MISCELLANEOUS

milliequivalents (meq)	≡ milliliters x normality
ppm ÷ 1000 x equivalent weight	≡ milliequivalents per gram
ppm ÷ 10 equivalent weights	≡ milliequivalents per 100 grams
milliequivalents per gram x 100	≡ milliequivalents per 100 grams
1 M (molar)	≡ 1 atomic weight (grams) [or mole] dissolved per liter water. Example: A 1M (one molar) solution of KNO_3 is 1 mole (101.103 grams) KNO_3 dissolved in 1 liter water. The mixture is brought up to volume after adding the solute (KNO_3)
1 mole (mol)	≡ the amount of substance of a system that contains as many elementary entities as there are atoms in 12 grams of carbon-12. The amount of substance containing Avogadro's number of molecules or formula units
Avogadro's number	≡ 6.022 1415 x 10^{23}
1 molal	≡ 1 atomic weight (in grams) dissolved in 1000 g solvent (divided by percent)

1 N (normal)	≡ 1 equivalent weight dissolved per liter water (or: 1 mol/L)
ppm ÷ by 10 000	≡ % (percent, or per centum)
percent **x** 10 000	≡ parts per million
1 percent	≡ 10 grams per liter (or kilogram)
1 percent	≡ 1.33 ounces (by weight) per 1 gallon water
1 percent	≡ 8.34 pounds per 100 gallons water
1 ream (paper)	≡ 500 sheets
1 case (paper)	≡ 2800 sheets
1 exasecond	approximately 32 billion years
1 exameter	approximately 110 light years
1 light year	≈ 5 865 696 000 000 miles ≈ 9 460 800 000 000 kilometers
speed of Light	≡ 2.998 x 10^8 meters per second ≡ 186 282 miles per second
acceleration of gravity	≡ 9.806 meters per second squared
pounds per acre x 1.12	≡ kilograms per hectare
kilograms per hectare x 0.893	≡ pounds per acre

SOME COMMON FORMULAS

volume of sphere	$V \equiv 4\pi r^3$ divided by 3, in which π is 3.1416 and r the radius [also, diameter^3 x 0.5236]
volume of cube	$V \equiv a^3$, in which a is one of the edges
volume of cone	$V \equiv 1/3\pi r^2 h$, in which π is 3.1416,

	r the radius of the base, and h the height [also: area of base x 1/3 height]
volume of rectangular prism	$V \equiv abc$, in which a is the length, b is the width, and c is the depth
volume of pyramid	$V \equiv 1/3$ area of the base x height $V \equiv 1/3\, Ah$, (in which A is the area of the base and h the height)
volume of cylinder	$V \equiv \pi r^2 h$, in which π is 3.1416, r the radius of the base, and h the height
diameter of circle	$D \equiv$ circumference x 0.318 31
area of circle	$A \equiv \pi r^2$ in which π is 3.1416 and r the radius [also, diameter^2 x 0.7854]
area of triangle	$A \equiv$ half the base x height
area of square	$A \equiv a^2$, in which a is one of the sides
area of rectangle	$A \equiv ab$, in which a is the base and b the height
area of trapezoid	$A \equiv h(a + b)$ divided by 2, in which h is the height, a the longer parallel side, and b the shorter
area of regular pentagon	$A \equiv 1.720 a^2$, in which a is one of the sides
area of regular hexagon	$A \equiv 2.598 a^2$, in which a is one of the sides
area of octagon	$A \equiv 4.828 a^2$, in which a is one of the sides
area of ellipse	$A \equiv$ the product of both diameters x 0.7854
circumference of circle	$C \equiv \pi d$, in which π (pi) is 3.1416 and d is the diameter
(ordinary) frequency	$\nu = \omega / 2\pi$

RATES OF APPLICATION

1 ounce per square foot	≡ 2722.5 pounds per acre
1 ounce per square yard	≡ 302.5 pounds per acre
1 ounce per 100 square feet	≡ 27.2 pounds per acre
3.5 ounces per 100 square feet	≡ 100 pounds per acre
7.5 ounces per 100 square feet	≡ 200 pounds per acre
8.75 ounces per 100 square feet	≡ 250 pounds per acre
14.75 ounces per 100 square feet	≡ 400 pounds per acre
1 pound per 100 square feet	≡ 435.6 pounds per acre
2 pounds 5 ounces per 100 square feet	≡ 1000 pounds per acre
4 pounds 10 ounces per 100 square feet	≡ 2000 pounds per acre
1 pound per 1000 square feet	≡ 43.6 pounds per acre
1 gallon per acre	≡ 1/3 ounce per 1000 square feet
5 gallons per acre	≡ 1 pint per 100 square feet
100 gallons per acre	≡ 2.3 gallons per 1000 square feet ≡ 1 quart per 100 square feet
100 pounds per acre	≡ 2.3 pounds per 1000 square feet ≡ 0.2296 pound per 100 square feet
1 ton per acre	≡ 20.8 grams per square foot
1 ton per acre	≡ 1 pound per 21.78 square feet
1 ton per acre	≡ 25.12 quintals per hectare
1 gram per square foot	≡ 96 pounds per acre
1 ounce per square foot	≡ 2722.5 pounds per acre

1 pound per acre	≡ 0.0104 gram per square foot ≡ 1.12 kilograms per hectare
pounds per acre	≡ grams per square foot x 96
tons per acre	≡ kilograms per 48 square feet
pounds per square feet **x** 21.78	≡ tons per acre
pounds per square feet **x** 43,560	≡ pounds per acre

POTASSIUM AND PHOSPHORUS

% P (in a fertilizer)	≡ % P_2O_5 x 0.43
% P_2O_5 (in a fertilizer)	≡ % P x 2.29
% K (in a fertilizer)	≡ K_2O divided by 1.2
% K_2O (in a fertilizer)	≡ % K x 1.2

APPROXIMATE MEASURES

1 level teaspoon	≈ 1/6 ounce
1 level tablespoon	≈ 1/2 ounce
1 level cup	≈ 8 ounces
1 pint	≈ 1 pound fertilizer
1 quart	≈ 2 pounds fertilizer
1 gallon	≈ 8 pounds fertilizer